Toward an
Evolutionary Regime for
Spectrum Governance

Toward an Evolutionary Regime for Spectrum Governance

Licensing or Unrestricted Entry?

WILLIAM J. BAUMOL
DOROTHY ROBYN

AEI-BROOKINGS JOINT CENTER FOR
REGULATORY STUDIES
Washington, D.C.

Toward an Evolutionary Regime for Spectrum Governance may be ordered from:
Brookings Institution Press
1775 Massachusetts Avenue, N.W.
Washington, D.C. 20036
1-800/537-5487 or 410/516-6956
E-mail: hfscustserv@press.jhu.edu
www.brookings.edu
All rights reserved

Library of Congress Cataloging-in-Publication data
Baumol, William J.
    Toward an evolutionary regime for spectrum governance : licensing or unrestricted entry? / William J. Baumol, Dorothy Robyn.
    p.   cm.
    Summary: "Evaluates two options for spectrum governance—a tradable license approach and a commons approach—in terms of interference, investment in innovation, monopoly power, diversity, rural service, and vested interests versus adaptability"—Provided by publisher.
    Includes bibliographical references and index.
    ISBN-13: 978-0-8157-0849-0 (pbk. : alk. paper)
    ISBN-10: 0-8157-0849-1 (pbk. : alk. paper)
    1. Radio frequency allocation. 2. Telecommunication policy. I. Robyn, Dorothy L. II. Title.
    HE8675.B38 2006
    384.54'524—dc22   2005027493

9 8 7 6 5 4 3 2 1

The paper used in this publication meets minimum requirements of the American National Standard for Information Science—Permanence of Paper for Printed Library Materials: ANSI Z39.48-1992.

Typeset in Adobe Garamond

Composition by OSP, Inc.
Arlington, Virginia

Printed by R. R. Donnelley
Harrisonburg, Virginia

Contents

Foreword

THIS VOLUME IS ONE in a series commissioned by the AEI-Brookings Joint Center for Regulatory Studies to contribute to the continuing debate over regulatory reform. The series addresses several fundamental issues in regulation, including the design of effective reforms, the impact of proposed reforms on the public, and the political and institutional forces that affect reform.

In this study, William Baumol and Dorothy Robyn take a fresh look at how radio frequency spectrum should be governed, in response to recent calls for the Federal Communications Commission (FCC) to eschew licensing in favor of a spectrum "commons." Spectrum is critical for a number of sectors in our economy, including telecommunications and information technology. The authors, like economists going back to Ronald Coase, maintain that the FCC's traditional, administrative approach to spectrum governance has been wasteful and perverse. They compare two alternative approaches—reliance on the market to allocate flexible, spectrum-use licenses and an unlicensed ("commons") approach that would rely on technology to limit interference—in terms of the six key issues with which a rational spectrum regime must deal (interference, investment in innovation, monopoly power, preservation of diversity, service to rural areas, and the tension between vested interests and the need for adaptable arrangements). They conclude that while a market approach to spectrum governance is not perfect, a commons approach has severe shortcomings. However, they stop short of embracing the pure market regime that some economists favor, instead emphasizing the need to preserve the government's ability to adapt the spectrum regime to accommodate currently unforeseeable changes in technology and other circumstances.

Like all Joint Center publications, this monograph can be freely downloaded at www.aei-brookings.org. We encourage educators to distribute these materials to their students.

ROBERT W. HAHN

ROBERT E. LITAN

AEI-Brookings Joint Center for Regulatory Studies

Acknowledgments

THIS MONOGRAPH WAS WRITTEN under the sponsorship of QUALCOMM Incorporated. It must be emphasized that there was no attempt to induce the authors to accept any corporate position and that the authors take full responsibility for all of the opinions and observations offered here. They would, however, like to acknowledge the very valuable suggestions and assistance received from William Zarakas, Evan Kwerel, Gregory Rosston, Thomas Hazlett, Robert Hahn, Gerald Faulhaber, and Ellen Goodman.

1 Toward an Evolutionary Regime for Spectrum Governance

Supposing that the earth yielded spontaneously all that is now produced by cultivation; still without the institution of property it could not be enjoyed; the fruit would be gathered before it was ripe; animals killed before they came to maturity; for who would protect what is not his own? There would be a strange mixture of plenty, waste, and famine.

In this country, for instance, where the only common property consists in hedge-nuts and blackberries, how seldom are they allowed to ripen?

Jane Marcet, *Conversations in Political Economy*, 1819[1]

I think there is a world market for maybe five computers.
Attributed to Thomas Watson, Sr., chairman of IBM, 1943

TODAY, THE RADIO FREQUENCY spectrum is the shared resource that perhaps most strikingly and most pervasively affects the well

---

1. Marcet (1819, pp. 60–61).

being of society.[2] Its use is governed by a set of rules and narrow restrictions, designed to limit interference, whose origins go back nearly a century. While in recent years some of those rules have been replaced by more flexible, market-like arrangements, the fundamental approach of this institution remains essentially unchanged.

There is widespread agreement that the current institutional arrangements are a source of major inefficiency and waste, and that the public interest calls urgently for some substantial modifications. So far as we are aware there are no articulate defenders of the current regime, and a number of substantially different substitutes have been proposed as markedly superior in terms of their consequences for the general welfare.

Here it will be agreed that there is substantial room for improvement. However, it is important to emphasize that whatever arrangement is shown to be better today, *it is not because our innate ability to understand the issues is superior* to that of our forebears who designed today's rules. Rather, the primary source of any current shortcomings is that circumstances, notably in the form of the range of products that make use of the spectrum, as well as the associated technology, have changed in ways that were

2. A number of the discussants of this issue have argued that the spectrum is a misleading fiction and that what really is at issue is the set of wireless transmissions, the equipment that generates and receives them, and the possibility that they may interfere with one another. Here there is no need for us to take a position on the existence of an entity that can be labeled "the spectrum." This matter does not affect the substance of the discussion that follows, though it will at times be convenient to refer to the spectrum as though it were a real and well-described entity.

not *and could not have been* foreseen at the time the current procedures were adopted.

We cannot claim to have become better at foretelling the future than our predecessors were. The wisest and best informed of us have repeatedly gone astray when we have attempted fortune telling. Rather, in our view, the only thing we can confidently assert about the future is that it will surprise us.

In our judgment, understanding of this reality is critical for the design of a modified spectrum governance regime that will deserve to endure because of its continued public benefits. We ought to avoid any substitute rules that, like those currently in place, establish a powerful vested interest in prevention of change and thereby make any necessary or desirable change difficult to institute and highly contentious. Such rules are very likely at some point in the future to become as counterproductive as today's arrangements.

The implication is that any good rules will have to be readily changeable as demands, uses, technology, and other critical determinants evolve in unpredictable ways. This is surely one of the most critical criteria for the design of programs for allocation and utilization of the spectrum, and it is an issue that, at best, has not been addressed as more than a peripheral matter in the discussions of proposals for revised spectrum policy.

Rather, redesign of the current spectrum governance regime has been discussed in terms of the choice between two fundamentally different alternatives. One is a market approach that treats licensed access to the spectrum as private property. The other approach would eliminate licensing altogether, at least for a portion of the spectrum. By analogy with medieval practice for a reserved portion of the available land, the latter proposal would treat parts of the spectrum as common property, essentially open to access by any and all who desire to use it.

As this paper shows, neither the current regime nor the commons approach can serve the public interest well. On the contrary, they both can be expected to result in misuse, inefficiency, and waste. Rather, the welfare of the community will be best promoted by an intermediate arrangement—a variant of the market approach that relies on property rights of unlimited duration, controls spectrum use by way of a market mechanism, materially eases but does not eliminate restrictions upon the uses to which rights can be put by their proprietor, and permits regulatory intervention in a limited number of situations where a pure market approach can damage economic efficiency. Underlying this public interest approach is recognition of the market mechanism's powerful contribution to efficient resource utilization, balanced by the urgency of not overlooking the well-known imperfections to which a market regime is vulnerable. But even that market regime, if its benefits are to long endure, must have built into it effective and workable provisions for easy adaptation to changing circumstances.

In sum, what is emphasized here are the benefits to the public of reliance on the market, provided that appropriate steps are taken to deal with its shortcomings.

It will also be shown that the choice between a market proprietary regime and a commons approach is not irrelevant for the issue of malleability of the rules. If not appropriately constrained, both approaches can easily give rise to those vested interests that are the primary obstacles to evolution and adaptation of rules to changing circumstances. However, it will be argued that it is the market regime, if properly designed, that is more amenable to adaptability.[3]

Advocacy of a regime of property rights and markets may well be attributed to a common professional predisposition of economists, including the current authors, to envision the market mechanism as the preferable and effective solution to many problems of the economy. However, that is somewhat of a misunderstanding. The economic literature emphasizes that the market has many imperfections, along with its benefits. As is emphasized below, there is no perfect solution to problems such as those posed by utilization of the spectrum, and rational choice among the available options entails careful weighing and balancing of the differing shortcomings of the available alternatives. That is the objective of this monograph.

3. At this point it may be appropriate to emphasize that the authors can claim no qualifications as engineers or scientists knowledgeable about the complexities of spectrum usage and the associated technology. At least in this arena we do have a good deal to be modest about. Consequently, any statements in this report about such matters should necessarily be interpreted with caution. However, one may hope that, as in the emperor's new clothes, lack of technical sophistication can contribute somewhat to clarity of vision regarding the basic issues and their essence.

2 Traditional and Alternative Approaches to Spectrum Governance

THE ORIGINS OF THE TRADITIONAL approach to spectrum governance go back more than eighty years, to when the federal government began assigning exclusive rights to use individual blocks of bandwidth as a way to limit overcrowding and interference.[1] In keeping with the skepticism with which markets were viewed at the time, the Federal Communications Commission (FCC) pur-

1. The U.S. Department of Commerce had previously enforced (and several courts had recognized) a de facto property right, under which the first user effectively owned the spectrum. But in 1926, following two unfavorable court decisions, Secretary of Commerce Herbert Hoover stopped enforcing this "priority in use" standard. Predictably, new entrants began broadcasting on the frequencies of popular radio stations, causing signal interference and chaos in major radio markets, and prompting Congress in 1927 to establish the Federal Radio Commission for the purpose of managing use of the radio spectrum. The Communications Act of 1934 transferred that agency's authority to the newly created Federal Communications Commission. See Hazlett (2001, pp. 350–57). For detailed accounts of the history of radio and the origins of radio regulation, see Hazlett (1990) and Benkler (1998).

sued a command-and-control approach to spectrum licensing that became a textbook example of the unintended consequences of regulation. First, the FCC allocated frequencies to particular uses and permitted little significant modification in the officially designated use either by existing licensees or by new entrants. These restrictions prevented spectrum from being put to more constructive uses, and over time they inhibited innovation. Second, the commission severely limited the transferability of the licenses, so that opportunities for profitable resale or leasing were either eliminated or narrowly circumscribed. Moreover, because profitable resale or rental was precluded, the holder of a license had little or no incentive to avoid hoarding of any portion of the spectrum assigned to him but for which he had no current use. Finally, the FCC's comparative hearing process used to select among competing license applications was costly and slow, delaying applicants' ability to respond to evolving market conditions and often requiring them to reveal proprietary business ideas.[2]

Although the FCC has begun to dismantle this long obsolete governance regime, much if not most of the spectrum that falls within the commission's jurisdiction is still covered by administrative rules and restrictions.[3] And on broader policy decisions, the FCC remains vulnerable to political pressure from incumbent licensees and other entrenched interests. (One FCC chairman joked that the agency's initials stood for "firmly captured by cor-

2. For a concise statement of this and other problems with FCC regulation, see Federal Communications Commission (2001).

3. Federal Communications Commission (2002).

4. Reed Hundt, speech given at the Center for National Policy, Washington, D.C., May 6, 1996.

porations."[4]) As one illustration of this, television broadcasters famously continue to occupy largely idle "beachfront" spectrum despite digital entrepreneurs' pleas for bandwidth. More generally, experts claim that much of the spectrum is not in use at any given time. This underutilization of the airwaves by license holders, at a time when there is said to be a "spectrum drought," reflects, at least in part, the continuing rigidity of the FCC's legacy governance regime.

Market Approach

Economists, who have long been skeptical about the ability of government agencies to allocate resources efficiently by "picking winners," have preponderantly favored a market approach to the allocation of resources generally, and to the allocation of the spectrum in particular. As early as 1959, Ronald Coase wrote that spectrum was a fixed factor of production, like land or labor, and should be treated in the same way, with its use determined by the pricing system and awarded to the highest bidder. Coase concluded that government allocation of spectrum-use rights was not necessary to prevent interference and that, in fact, by preempting market allocation of spectrum, regulation was the source of extreme inefficiency.[5]

Economists since Coase have favored a market-based approach

5. Coase (1959). Coase's Nobel prize–winning essay the following year, "The Problem of Social Cost" (Coase 1960), followed directly from his work on the FCC.

to spectrum governance.[6] This approach is characterized by three key elements (most of which Coase identified):

—well-defined, exclusive rights to the use of spectrum—that is, assignment of the exclusive right to use a particular frequency in a specific geographic area, subject to technical rules designed to limit interference between licensees;[7]

—a market-type mechanism such as an auction for an initial allocation of spectrum rights and a secondary market in which those rights can be shared or transferred through sale or lease; and

—allowance for maximum flexibility as to the types of commercial services that can be provided.

Such an approach has two important advantages over administrative licensing. The first is what might be thought of as efficiency in use. Economists believe that the profit motive will deliver spectrum, like any other valuable resource, to those who can put it to the uses most desired by the public. Specifically, markets will excel at the task of spectrum allocation because they are driven by those who have most immediate knowledge of and experience in spectrum use. Moreover, efficiency of use is

6. See, for example, De Vany and others (1969), Hazlett (1998, 2001, 2005), Levin (1971), Minasian (1975), Rosston and Steinberg (1997), Shelanski and Huber (1998), Spiller and Cardilli (1999), and White (2000).

7. The three dimensions of a "parcel" of spectrum are frequency, space, and time; the time dimension is less important because most licenses provide for twenty-four-hour-a-day usage. In addition, a licensee's interference rights are a function of technical rules. Traditionally, these rules have taken the form of technical specifications (for example, location and direction of transmitters) that could not be changed without approval. In certain bands, the FCC has replaced technical specifications with power limits, so that licensees have greater freedom as to how they operate within those limits. See Kwerel and Williams (2002, pp. 42–44) and Faulhaber (forthcoming, pp. 31–32).

enhanced by the market mechanism because markets exact automatic financial penalties from any user who employs resources wastefully or inefficiently. The market mechanism also penalizes unproductive hoarding of spectrum by forcing the holder to incur any required carrying cost without any offsetting revenues.

Over time, the inexorable pressure to make efficient use of a scarce resource such as spectrum leads to increased investment and innovation—a second important advantage that markets provide relative to administrative licensing. In the absence of the FCC's rules and cumbersome decisionmaking process, a licensee would have greater incentive to upgrade equipment so as to use less spectrum capacity, and an entrepreneur who wanted to offer a new service merely would have to bid for the needed spectrum.

The FCC has begun to introduce market mechanisms in recent years; in particular, since Congress authorized their use in 1993, the commission has held auctions to award most new spectrum licenses. Auctions are a major improvement over administrative assignment of licenses. However, only about 7 percent of the most valuable spectrum is available for market allocation.[8]

Commons Approach

Ironically, not long after the FCC finally heeded economists' advice and introduced the use of auctions to allocate certain licenses, a small group of technologists and legal scholars, among others, began urging the FCC to eschew licensing altogether and

8. Kwerel and Williams (2002, p. 1).

instead treat some or all of the spectrum as a common resource, or commons, in which there would be no limits placed on the number of users.[9] There was some precedent for a commons approach at the time, and since then the FCC has designated additional frequencies for unlicensed use. In part as a result of the perception that this limited experience with unlicensed spectrum had been successful, support for a commons (or unlicensed) approach to spectrum governance has grown.

Proponents of a commons approach base their case primarily on recent and anticipated developments in technology. These include spread-spectrum devices that trade off power for bandwidth, "spreading" a weak signal across a wide swath of frequencies, and "smart" radio devices designed to seek out what we may think of as temporarily unused pockets in the spectrum, so that by leaping from one temporarily unused region to another, the activity in question can proceed with little or no danger of its causing crowding or interference. Commons proponents maintain that the "amount" of the spectrum that is occupied by many particular uses is now minimal, and that there are ways in which the interference created by any particular use can be reduced well below what was previously possible.

9. See, for example, Benkler (1998 and 2002), Lessig (2001), and Werbach (2004). Proponents of a commons approach appear to be as critical of administrative licensing as are economists, but for a different reason: whereas economists oppose the use of *regulation* to allocate licenses and restrict their use, commons proponents oppose *licensing*—that is, the *exclusive assignment of frequencies*. From their perspective, moreover, the use of a market approach to allocation of licenses represents little if any improvement on administrative licensing.

The bottom line in all this is the claim that this new technology makes licensing—that is, the exclusive assignment of rights to use spectrum—both unnecessary and impractical. It is unnecessary because, far from being in seriously short supply, there is now an abundance of capacity in at least certain portions of the spectrum. Thus there is little or no need for the imposition of exclusivity to prevent overuse or uses that cannot avoid serious interference with one another. Licensing is said to be impractical because the traditional approach that partitions spectrum into narrow bands cannot easily accommodate the newer technologies that allow signals to leap from band to band (smart radio) or that transmit an extremely weak signal across a wide range of frequencies (ultrawideband). Proponents claim that the transaction costs of arranging to get the necessary spectrum rights for this newer technology—whether in a spectrum market or through an FCC regulatory process—would be prohibitive. Those who offer these observations conclude that the provision for licensing of spectrum rights, built on the premise that overcrowding and interference urgently require the imposition of exclusivity to prevent those undesirable results, is now far more restrictive than the situation warrants, if not entirely excessive.

Although the proponents of a commons regime reject the exclusive assignment of spectrum rights that is essential to the market approach, they maintain that a commons approach would nevertheless harness market forces by facilitating a market for end-user equipment. The analogy is to the U.S. system of roads and highways, which by providing "free" access has facilitated a vibrant automobile market. In the view of commons proponents, manufacturers of end-user equipment operating in a commons regime would have an incentive to use spectrum efficiently (efficiency in use) and to invest in innovation (dynamic efficiency) that would not be demonstrably inferior to the incentives motivating licensees in a market regime.

Easing of the requirement for exclusive assignment of spectrum rights, it is claimed, can benefit the consuming public by enhancing the variety of spectrum services consumers are offered and by increasing the quality of these services and reducing their costs, including the costs associated with exclusive licensing. Proponents of this approach assert there is an opportunity to attain a marked reduction of government interference in the economic activities that make use of the spectrum and to do this without giving this invaluable resource over to particularly favored private interests. In this view, it is only the vested interests created by the old regime that can suffer any disadvantage.

It is also suggested that a commons approach will deal effectively with two possibly serious problems that can beset spectrum use: the threat of monopoly power and the elimination of diversity. First, unrestricted entry will curb monopolistic behavior because overcharging and excess profits will quickly attract new entrants who will seek to obtain a share of the gains, thereby introducing the competition that will curtail any such exercise of monopoly power. Second, unrestricted entry will encourage diversity of spectrum uses and users by offering access to communicators holding every point of view, to diverse technological approaches, to different art forms, and the like, evidently a significant desideratum for the utilization of the spectrum.

3 *Key Issues for a Rational Spectrum Regime*

TO EVALUATE THE TWO alternatives to current spectrum policy discussed above, we turn next to a list of considerations that must be taken into account in designing a fully defensible set of spectrum utilization arrangements. Specifically, we compare (a variant of) the market regime and the commons regime in terms of the six key problems with which a rational spectrum regime must deal:

—control of prospective interference, a special case of a critical and widespread economic problem that economists call externalities;

—encouragement of investment in innovation;

—prevention of monopoly power or its exercise that may occur if one proprietor or group of proprietors achieves control of too large a share of spectrum capacity;

—preservation of diversity, such as is threatened when an absolutist government seizes control of broadcasting;

—recognition of the widespread desire to encourage provision of broadband service to rural areas; and

—preservation of adaptability to evolving circumstances such as advancing technology and changing consumer needs and preferences.

In the course of analyzing these six issues, we will offer specific recommendations for appropriate policy design.

Interference and the General Issue of Externalities

The essence of the spectrum-usage problem is the phenomenon of crowding, which arises because the spectrum, at least for now, remains a scarce resource and is therefore subject to costly mutual interference by the profusion of users. Economists, who view the unregulated market as a valuable but fallible instrument for direction of the economy, recognize this phenomenon as one of the most pervasive and significant of the shortcomings that beset the market mechanism. It is therefore appropriate to examine this one subject with some care.

Many economic activities generate incidental benefits or incidental harm to others for whom they are not specifically intended.[1] The market deals poorly with *incidental and unintended* side effects because such external effects, or externalities, do

1. For example, a scientist who achieves a research breakthrough that he or she can patent and use to acquire a financial reward also often provides unintended benefits to researchers in other fields who learn thereby how to perform their research and development more effectively. Thus the first researcher unintentionally provides benefits to others but receives no payment in return for this part of his or her contribution. Economists say that this

not enter into the profit and loss calculation of the businesses and other economic actors who create them.

Externalities cause the price system to misallocate resources because the market system achieves efficiency by rewarding producers who serve consumers well and at the lowest possible financial cost. This argument breaks down, however, as soon as some of the costs and benefits of economic activities are left out of the profit calculation. When a firm pollutes a river and thereby reduces its oxygen content, it uses up some of society's resources just as surely as when the firm burns coal. However, if the firm pays for coal but not for the use of oxygen, we can expect the firm's management to be economical in its use of coal and wasteful in its use of oxygen.

In an important sense, the source of the market mechanism's difficulty here lies in society's rules about property rights. Coal mines are *private property;* their owners will not let anyone take coal without paying for it. Thus coal is costly and so is not used wastefully. But waterways are not private property. Because they belong to everyone in general, they belong to no one in particular. Therefore, anyone can use waterways as free dumping grounds for wastes that spew poisons into the water and use up the water's oxygen that is vital for underwater life. Because no one pays for

---

activity generates a beneficial externality. That is, the activity creates benefits that are external to, or outside, the intentions and interests of those who are directly involved in the activity. Similarly, some activities incidentally and unintentionally impose costs on others. For example, the owners of a motorcycle repair shop may create a lot of noise for which they pay no compensation to their neighbors. Economists say these owners produce a detrimental externality. Pollution is the classic illustration of a detrimental externality.

the use of the socially valuable dissolved oxygen in a public waterway, people will use that oxygen without caution. The fact that waterways are exempted from the market's normal control procedures via pricing is therefore the source of a detrimental externality.

Where a firm's activity causes detrimental externalities, smaller outputs than those that maximize profits will be socially desirable. This is so because private enterprise has no motivation to take into account any costs to others for which it does not have to pay. In fact, competition *forces* firms to produce at as low a private cost as possible because if they do not, rivals will be able to take their customers away. Thus competition *compels* firms to make socially excessive use of resources for which they are not required to pay or pay fully. As a result, goods that cause detrimental externalities will be produced in undesirably large amounts because they have social costs that are not paid by the supplier firms.

The implication for possible overuse of the spectrum as a resource that is provided "for free" should be clear. In the absence of some mechanism that brings about an internalization of the social costs, where one spectrum use threatens damaging interference with other spectrum uses, overcrowding can be expected.

Use of Market Mechanisms to Control Interference

Although the market, acting on its own, does nothing to cure externality problems, the market mechanism does offer several effective ways of dealing with such difficulties. These mechanisms entail some government intervention, but they leave matters to the market once government has adopted appropriate modifications of the rules of the game. Three powerful market mechanisms have been used to address environmental externalities: private property rights with rights of negotiation, government taxation of damaging externalities, and government provision of tradable permits.

PRIVATE PROPERTY RIGHTS WITH RIGHTS OF NEGOTIA-TION. Private property rights with rights of negotiation is the approach associated with the work of Ronald Coase.[2] Coase's greatest insight was to see that once well-defined property rights are assigned to a resource that is subject to overuse or crowding, the parties involved—the owner, the creators of the damaging externalities, and those who bear the consequences—have an incentive to reach an efficient (that is, mutually beneficial) settlement. Previously, economists had written as though such an outcome was possible only through direct government regulation of the externality (for example, requiring smokestack scrubbers) or by imposition of a tax on the creation of damaging externalities (such as an emissions tax), as discussed below. Coase showed that neither regulation nor taxes are needed to attain the efficient outcome as long as the affected parties can readily enter and enforce contracts that are mutually beneficial.[3]

2. See Coase (1959, 1960).

3. Consider a public pond used for fishing, which if uncontrolled, is prone to overuse and depletion of the fish stock to a degree that prevents the fish population from maintaining itself. If the pond becomes a private property, the owner can be expected to take steps to prevent depletion of the valuable fish stock. The proprietor may, for example, fence the pond and place a charge upon fishing activity with a price per fish removed from the pond. If monitoring and collection become too costly, the proprietor may even contract with a few commercial fish suppliers, setting terms that include a negotiated price and a quota on the number of fish that can be taken. The

Coase's method will work where the number of parties involved, including the creators of the damaging externalities and those who bear their consequences, is limited so that direct negotiations are feasible and the determination of rational transactions terms is a practical undertaking. Where a larger number of parties is involved or where the identity of the parties now and in the future is not even known, this approach can become unworkable, as Coase was careful to point out. Coase first set out these ideas in his 1959 analysis of FCC regulation.[4] He concluded that well-defined property rights in spectrum—rights that could be transferred in whole or in part—would go a long way toward solving the problem of radio signal interference. He acknowledged, however, that if the number of parties involved in certain spectrum transactions proved to be unduly large, regulation might be necessary—but only as a supplement to transferable property rights.

GOVERNMENT TAXATION OF DAMAGING EXTERNALITIES. A second market mechanism (and the one economists had predominantly advocated before Coase's article appeared) relies on government taxation of damaging externalities, as is done by the much debated tax on carbon emissions. In the current case, the analog would be a tax on spectrum emissions. However, taxes on externalities are never politically popular, for reasons that are obvious enough, even though where they have been employed, they are

---

owner's self-interest will lead to terms that maximize the long-term value of the fish stock and the pond even beyond the proprietor's lifetime, because that will also maximize the current market value of the asset, whether retained or resold.

4. Coase (1959).

reported to have proven effective. There is a second reason why they may not be workable in management of spectrum use. Taxes on emission of damaging externalities require the ability to monitor and measure the amount of the undesirable effects upon others created by each individual emitting source, because to provide the appropriate disincentive for creation of the externality, the size of the tax bill should be proportional to the damage the emitter produces. But such measurement of the interference externality is probably very difficult with, if not entirely beyond, current technical capabilities. Alternatively, the regulator needs to be able to predict what price (tax) will result in the optimal level of spectrum emissions. But that, too, is a difficult exercise, fraught with uncertainty. If the tax is set too low, it will lead to a harmful level of interference. If the tax is set too high, it will impose an opportunity cost by discouraging licensees from making efficient use of the spectrum. Although approximations are possible, use of the tax mechanism to control spectrum interference is difficult and risky—one reason the FCC is more comfortable relying on emissions limits.

TRADABLE PERMITS. The tradable-permit mechanism involves the provision by a government agency of a set of licenses that allows their holders to carry on their crowding activities, but not to exceed the amount of such activity specified in the permit. The agency may either sell the permit or give it away without charge. The permit holder in turn can resell the permit to others, either in whole or in part, and for any time period within that covered by the permit. If the number of permits issued by the government agency in charge is suitably limited, it is evident that the volume of entry will be restricted directly, and the agency can then seek to determine on the basis of the available evidence the amount of entry that can be permitted without inviting an unacceptable amount of crowding.

If the permits are transferable, and particularly if, as experience suggests, an organized market in the permits subsequently develops, there is a second and important way in which this arrangement can function: as a financial disincentive to excessive crowding of the resource. The point is immediately obvious if the permits are distributed via an auction process or if the government adopts another way to charge a price for each permit. But that is not the end of the story. The price that is determined on the organized permit market becomes a fee that is a disincentive for resource use by the prospective permit purchaser on that market. These prices then become costs to the resource-using firms that can be the same as they would have been required to pay if the government had imposed a substantial tax upon resource use.

Such a system of transferable permits is now in use in the United States for some types of pollutant externalities, and there now exist organized markets in which firms can and do buy and sell emissions permits. The price in these markets is, as usual, determined by supply and demand, so if the government issues very few such permits, the supply-demand mechanism will make their price high, and the result will be a correspondingly strong disincentive against polluting.

The Virtues of Tradable Licenses

The FCC's deregulation of the market for commercial mobile radio services (CMRS) licenses (the flexible licenses in the 800 MHz, 900 MHz, and 1.9 GHz bands that were assigned via auctions) represents a variant of the first and third approaches described above. To a considerable extent the FCC relies on a market in flexible, transferable property rights to control interference in the CMRS spectrum, with a great many interference problems apparently resolved through direct negotiations among licensees, as Coase envisioned. But because direct negotiations are not always feasible or efficient, the FCC supplements this market with other rules to control CMRS interference (for example, limits on the height and radiating power of antennae). Thus the FCC's approach is a modification of the "true" Coase approach, which relies solely on the market.

The FCC's approach also can be seen as a variant of the tradable-permit system (a system that has itself been described as "Coasian" because it uses clear, marketable property rights to solve externality problems). Unlike emissions permits, which are a fungible commodity, each spectrum license represents a unique, exclusively assigned bundle of rights (although licenses for spectrum in the same geographic area and frequency band apparently can serve as near perfect substitutes, depending on the circumstances).[5] Nevertheless, the exclusive nature of spectrum rights does not affect the basic function served by a market in tradable permits.

By whatever name (below, we use the term "tradable license" regime), this basic approach is a relatively efficient and effective method of solving the externality problem associated with spec-

5. If spectrum interference were controlled through a market in spectrum emissions licenses, the analogy to tradable pollution-emissions licenses would be more direct. For an analysis of the potential for such a market, see De Vany and others (1969) and Minasian (1975).

trum use. The transferability of the licenses, and the emergence of a competitive forum in which supply and demand determines who will get licenses and at what price, harnesses the market mechanism for the job of interference control. A major advantage of tradable licenses is that they tend to be allocated to those who can make the most highly valued use of them. Since firms derive much of their funding from the capital market, and that market provides greater amounts to those with more promising investment prospects, it will be they who can make the highest bids for the licenses. Other advantages of this approach include the transparency with which initial licenses are awarded and the reduction, if not elimination, of opportunities for outside influence or rent seeking.

Unlicensed Spectrum Use and Interference

Leading proponents of a commons regime acknowledge that the market is superior to unconstrained entry as a method for allocating most physical resources. However, they maintain that spectrum is different because new technology makes it possible for users to add capacity and avoid interference dynamically, by means such as ability to find and utilize uncrowded portions of the spectrum.[6] But a closer look suggests that a policy of unlimited entry is likely to have the same detrimental effects upon spectrum usage that it has on usage of other shared resources.

To be sure, unlicensed use of spectrum has its place. Consumer devices such as microwave ovens and cordless phones traditionally have been unlicensed: the short range of transmission, combined

6. Werbach (2004).

with strict power limits, means that the use of these devices in one home does not create interference in another. And Wi-Fi technology has been a major commercial success, allowing consumers to download e-mail and tap into the Internet via unlicensed access. However, unlicensed transmissions currently use relatively undesirable portions of the spectrum (Wi-Fi spectrum was widely regarded as a "garbage band"); by contrast, the opportunity cost of the spectrum to which commons advocates seek to expand unlicensed access is very high. Moreover, Wi-Fi hotspots avoid congestion because the transmissions are extremely short range, and individual establishments can effectively restrict their own airspace, thereby exercising de facto property rights. Even so, Wi-Fi users in some urban areas reportedly are encountering major interference as a result of the sheer number of networks now in place.[7] Moreover, large Wi-Fi users such as corporations and universities generally purchase proprietary access to avoid such interference.

Interference is even more of a problem for wireless Internet service providers (WISPs) using unlicensed spectrum to serve rural locations and other broad geographic areas.[8] Although Intel is reportedly deploying new wireless technology (WiMAX, or wire-

7. Mike Musgrove, "Here, There, WiFi Anywhere: Wireless Web's Spread Is Crossing Our Signals," *Washington Post*, April 25, 2004, p. F1.

8. See, for example, Andrew Seybold, "Slogging through the Hype," *Wireless Week*, September 1, 2004 (www.wirelessweek.com/index.asp?layout=articlePrint&articleID=CA448732). A Google search turns up scores of articles and other documents on "WISP interference," including marketing material on a "Moot Court Manual" designed to advise competing WISPs on the feasibility of suing one another for unlicensed interference. See "Moot Court Manual—License Exempt Interference" (www.part-15.org/shop/p15/moot.asp [September 2005]).

less interoperability for microwave access) that will be able to span much greater distances than Wi-Fi, Intel appears to be positioning it primarily for use in licensed spectrum—further evidence that overcrowding and interference in the unlicensed spectrum are likely, at least for some time, to remain a chronic condition.[9]

Such interference is inevitable under a spectrum regime in which the market is not constrained by any restrictions that limit entry: in deciding whether or not to enter, each entrant takes into account only the consequences that bear upon him- or herself and disregards the effects upon others. As the tragedy of the commons parable would predict, the spectrum tends to end up in overcrowding and overuse. Although new technology has expanded the amount of spectrum capacity that is effectively available, demand has more than kept pace, and this is unlikely to change.[10]

9. In a recent online interview, James Johnson, vice president of Intel's Communications Group, said, "The other big difference between Wi-Fi and WiMAX—starting right away—is that we're going to use licensed spectrum to deliver WiMAX." See "WiMAX: Wireless Broadband for the World—An Interview with Jim Johnson" (www.intel.com/netcomms/columns/jimj105.htm [September 2005]).

10. A recent *Business Week* story on wireless technology describes a fictional vineyard in Sonoma Valley where in a few years each grape vine will be equipped with a wireless sensor that can track data such as heat, moisture, and nutrients in the soil. According to the same story, Nestle is already installing ice cream vending machines that send daily sales reports and notify drivers if they are running low on individual products. "Special Report—Wireless Wonders," *Business Week*, April 26, 2004 (www.businessweek.com/ magazine/toc/04_17/B38800417wireless.htm). The prospective applications of wireless sensors (which are expected to shrink to the size of dust), communications systems, identification devices, and other wireless technology are virtually unlimited.

In fact, FCC efforts to accommodate new and better unlicensed wireless technology repeatedly have resulted in overcrowding, according to Hazlett: "When unlicensed entry thrives, the characteristic pattern is that overcrowding ensues. The history of unlicensed entry is a chase up the dial: the 900 MHz ISM band became congested, leading the FCC to open up the 2.4 GHz unlicensed band, which became crowded in major markets, leading the Commission to open up 300 MHz for the U-NII 5 GHz band."[11]

Some proponents of a spectrum commons have argued that this portrayal exaggerates matters and that self-interest will drive users toward appropriate restriction of the external effects, with the aid of technology that is just over the horizon. The analogy used is that of overcrowded public highways whose delays lead some drivers voluntarily to switch to the use of smaller roadways in order to save time. It is true that this will prevent those particular drivers from further exacerbating the delays on the major highways, but a little thought and some careful analysis show that this can only be depended on to ameliorate the problems marginally and leave the bulk of the overcrowding as it was.[12]

We all know from bitter experience that overcrowded roads generally continue to be overcrowded and unclog themselves only when significant tolls are adopted or some other rationing device is put into place, and spectrum is not essentially different.[13]

11. Hazlett (2001, p. 429, footnotes omitted).

12. The argument, indeed, has the logic of the Yogi Berra remark about the restaurant that is so overcrowded that no one goes there anymore.

13. Anthony Downs's "principle of triple convergence," which explains why we cannot build our way out of roadway congestion by adding lanes,

Granted, the existence of smart radio devices that can coordinate with one another could help to limit such overcrowding. But even if such technology does become available, the very nature of unlicensed spectrum—no single entity has the incentive or ability to coordinate user behavior—means that users will not necessarily upgrade their equipment to take advantage of the new capability. Over time, this impediment to the upgrading of equipment also discourages investment in innovation.[14]

Precisely because technology alone is not enough to prevent overcrowding, proposals for a commons approach to spectrum governance also envisage the use of direct government controls—specifically, adoption of common technical standards and protocols for different spectrum bands. Proponents of this approach argue that these "light-handed" regulations would provide rules of

---

applies equally to spectrum. According to Downs (1992), if a congested road is expanded, allowing traffic to move faster, drivers will soon alter their behavior in response, converging on the expanded road from other routes they have been using to escape congestion, moving from off-peak to peak travel times, and even shifting from other modes such as buses and trains. Moreover, expanded roads attract new development, which generates additional traffic.

14. One illustration of this problem is land mobile radio (so-called private radio bands that are typically used for intrafirm radio communications such as taxi dispatching). Although this spectrum is licensed, it is shared by unrelated users. According to one spectrum expert, "Users are stuck with old, technically inefficient equipment. Why? Because none of them has the incentive to adopt new equipment on their own that would free up spectrum for use by others. Instead, they came to the FCC with a proposal to transition over twenty-seven years to equipment that was not quite state of the art at the time of their proposal" (Gregory L. Rosston, testimony before the U.S. Senate Commerce Committee, March 6, 2003).

the road and help create a vibrant market in (unlicensed) interference-avoiding devices.

However, two basic problems beset such direct controls, rendering them less effective than tradable licenses as a way to limit interference. First, direct controls tend to be inefficient and costly, most immediately because they usually entail a one-size-fits-all approach: even within a single band of spectrum, an interference standard that is effective for one use or set of circumstances may be quite unsuited to another. The frequent result will be an inefficient use of spectrum entailing either under- or overcrowding. Compared with direct controls, tradable licenses can achieve a given reduction of interference, presumably to what can be deemed "a tolerable level," at a far lower cost.[15]

Second, government regulations create an incentive for private parties to invest substantial effort and resources into inducing regulators to serve their interests. This results in a politicized regulatory process that is time consuming and expensive and fails to serve the general welfare—just what critics agree is harmful about the current approach to spectrum governance. Under a commons regime, the locus of lobbying would shift from licenses to protocols and equipment, but in other respects the process is unlikely

15. For example, statistical estimates for several pollution control programs indicate that the cost of doing the job through direct controls can easily be twice as high as under a tax on polluting emissions, which basically works just like the tradable-permit approach. Why should there be such a cost difference? Under direct controls, the task of reducing emissions is usually not apportioned among the emitting firms on the basis of ability to reduce their emissions cheaply and efficiently. But the price of licenses induces their acquisition by bidders who can put them to best use and can therefore best afford to pay for them.

to improve. Granted, a market approach does not eliminate altogether the need for government rules such as power limits, but licenses themselves are allocated by means of a (market) mechanism that is speedy and is generally impervious to influence.[16]

In sum, as Faulhaber points out in an excellent paper on spectrum economics, "light regulation" is not a real option.[17] Regulation has always brought with it an unenviable record of inefficiency, litigation, and irresistible rent-seeking opportunities that tempt interested parties to take advantage of and even to distort the rules for the purpose of personal gain. There is no reason to believe that regulation of spectrum usage would be able to improve on this performance and its threat of damage to the public interest. Of course, regulation is sometimes unavoidable as an instrument for protection of the public interest. But where, as in the case of spectrum usage, a workable alternative is available, there is a strong presumption that regulation should be minimized if not avoided altogether.

In fact, the prediction that a commons regime would contribute to the politicization of the regulatory process is borne out by several ongoing campaigns that were inspired by overcrowding in the unlicensed bands. First, some of the wireless Internet service providers plagued by interference problems have sought the equivalent of an exclusive license in unlicensed spectrum (what

16. A third problem with direct controls is that they rely on the criminal justice system for enforcement. By comparison, licensing costs are more automatic and certain. That said, a market-based system will no doubt require some enforcement to ensure that property rights are honored.

17. Faulhaber (forthcoming).

one called a WISP Homestead Policy).[18] Although the service providers' frustration is understandable, their request is a contradiction in terms. In addition, a number of private parties apparently are pressing Congress and the FCC to allocate additional spectrum for unlicensed use. Moreover, they are now targeting spectrum in the TV bands, which is far more valuable than the spectrum currently used for unlicensed transmissions.[19] These campaigns illustrate both the inherent drawbacks of unlicensed entry (commons) as a mechanism for spectrum allocation and the opportunities such an approach would afford for continued rent seeking.

Investment in Innovation

One of the issues that has recurred in the debate over the best regime for operation of the spectrum services is the incentive for investment in innovation. The explosion in variety, sophistication, and value of spectrum services is evidently a tribute to the inventiveness of the twentieth century, and it is clear that the public interest is most likely to be served by an arrangement that provides effective incentives for innovative activity. The profit mechanism has, indeed, produced an outpouring of innovation unparalleled in previous history. It has accomplished this by pro-

18. See, for example, John Scrivner (2004), as quoted in Hazlett (2005, pp. 266–67).

19. Heather F. Weaver, "New Congressional Caucus Calls for Some Unlicensed 700 MHz Spectrum," *RCR Wireless News*, July 20, 2005 (rcrnews.com/printwindow.cms?newsId=23466&pageType=news).

viding effective financial rewards to the successful inventor and by making innovation the prime weapon of business competition, particularly in the economy's high-tech industries. In the case of the spectrum, the mechanism is the incentive structure under which returns to innovative effort accrue to the holder of the property (the license) that benefits from this effort. If no one holds such property, that source of incentive for the cost, effort, and risk entailed in the innovation process will be undercut. It surely is foolhardy, at best, to spurn the market mechanism as the instrument for stimulating abundant and powerful innovation when experience, as well as analysis, confirms that alternative arrangements have fallen so far and so universally behind the market's performance in the arena of innovation.

Advocates of commons have, however, suggested that in a commons regime, the issue will automatically be dealt with in another way: they maintain that it will be in the interest of manufacturers of equipment for spectrum-using activities to invest in the design of technological solutions to the interference problems, as a means to expand their volume of business. Thus, according to their arguments, equipment manufacturers would be as powerful an engine for innovation under a commons regime as license holders would be under a market regime. But these arguments overlook the fact that this approach entails yet another externality problem. If manufacturer A's investment in research and development succeeds in reducing interference problems and thereby makes it possible to expand the volume of spectrum activity, only part of the resulting increase in business is likely to go to A. Much of it will also go, in the form of a beneficial externality, to equipment manufacturers B, C, and D, some of whom may even be direct rivals of A. Because a substantial portion of the benefits of A's research and development expenditures go to others, it will certainly not pay A to spend as much on research and development as the public interest requires. One cannot put off, and rely on equipment manufacturers to take on, the task of dealing adequately with investment in means to prevent interference.

A commons regime also is deemed by its proponents to be friendly to innovation because it does not require new entrants to pay for spectrum. However, it is simply not true that freedom of entry requires valuable resources to be given away without charge. On the contrary, such a giveaway approach deprives the economy of the assurance that the valuable resources will go to those who need them most urgently, as well as the other efficiency attributes for which the market mechanism is noted. In contrast, a spectrum-utilization arrangement based on auctioning of licenses is a regime of open entry that offers those advantages. That freedom of entry does not require that valuable resources be provided without any charge to those who want to use them is easily confirmed by observing the way in which free entry works in arenas in which it is evidently present. Surely, in no industry is entry less constrained than it is in farming. Yet to achieve such free entry, it has not been necessary for land, seeds, and fertilizer to be provided without charge. Entry into air passenger transportation is also unrestricted, indeed so much so that the resulting creation of new airlines threatens the viability of the incumbents. Yet it has not been found necessary to offer the entrant free aircraft or free landing slots to induce establishment of new firms. On the contrary, granting of free resources as a means to enhance entry would only invite waste of these resources and their employment by those not in a position to make the most efficient use of them.

Also relevant is the issue of market structure. The economic literature has over the years debated whether innovation is best served by widespread competition among myriad small-sized firms, driven by rivalry to strive for better products and processes, or at the other extreme, by monopoly, which possesses the resources to invest liberally in research and development and which is in the best position to collect the benefits its innovative efforts generate, rather than their being spilled over to others.

The evidence provides a somewhat more complex picture. First, it suggests that a high proportion of the breakthrough inventions that first appeared in quantity with the advent of the industrial revolution was contributed by independent inventors and entrepreneurs, who, unlike the large enterprises, were less likely to be put off by unconventional ideas and technology. But the data also indicate that the further development and improvement of the resulting novel technology and the bulk of the resulting applications were carried out by larger enterprises. However, these firms were usually not monopolists and were in fact driven by constant competitive pressure to invest unceasingly in research and development to avoid being outperformed by their equally innovating rivals. The implication is that although technical progress is encouraged by freedom of entry for new and innovative firms, the larger enterprises also play a crucial role.

Monopoly Power

As with other services, there is reason to be concerned about the possibility that some threat to competitiveness may emerge in use of the spectrum. Most obviously, in an auction of licenses, it is conceivable that a large enterprise with deep pockets can acquire a sufficiently large share of the available licenses to provide it with monopoly power. At first glance the licensing option may seem to offer the most significant anticompetitive threat. But that is misleading. Even a commons with unrestricted entry does not preclude the presence of a dominatingly large enterprise, and in either case, if no other sources of market power are present, a simple ceiling on market share can eliminate the problem.

It is more difficult to deal with the monopoly-power problem where the source is structural, based on scale economies or network externalities. Scale economies evidently confer a competitive advantage on larger firms and ultimately prevent the survival of smaller competitors, even when there has been no misbehavior by the surviving enterprises. But once freed from competition, a surviving firm can choose to exercise its resulting monopoly power in a manner damaging to the public interest.

The consequences of what have come to be called "network externalities" are virtually the same. A service is said to entail a *network externality* if the value of the service to any of its users is enhanced whenever there is an increase in the total number of users. An example is a word processing program, incompatible with rival programs, that users want to employ to exchange messages with one another. Here again, the supplier that starts off biggest, in terms of the number of initial customers, will have an advantage over smaller rivals that will lead to its growth in market share and further enhancement of its competitive advantage, just as is well recognized for the case of scale economies.

Scale economies and network externalities are not easy to deal with in any manner that can claim to be optimal in terms of the public interest. But two observations here are critical if the problem thereby posed for spectrum management is to be properly understood.

First, it must be recognized that any threat to competition is not solved automatically either by the grant of property rights via licenses or by creation of a commons. It is true that a commons without entry restrictions does invite invasion of a market by smaller entrants. But that does not enable the entrant to compete effectively against an entrenched rival who derives a material competitive advantage from scale economies or network externalities. Second, these problems are not associated uniquely with the spectrum. The antitrust laws and regulatory agencies can be expected to be just as effective in dealing with the issue, here as elsewhere.[20]

20. But there is an important *caveat* about some usual antitrust presumptions here. As discussed earlier, Coase's negotiated settlement approach to the control of externalities based on private property inherently calls for direct negotiations among the affected parties (Coase, 1960). Although coordinated decisionmaking is understandably considered suspect from an antitrust point of view, it has become clear from experience in the arenas of research, development, and utilization of intellectual property that explicit and organized coordination and agreement procedures, even where consenting decisions are arrived at by direct horizontal competitors, can be socially advantageous. The proprietors of complementary inventions can hasten the

In sum, while the dangers to competition that can arise in spectrum management are real, they are no less manageable by the antitrust rules than such problems elsewhere.

Preservation of Diversity

Somewhat related to the issue of monopoly is the problem of impediments to diversity, elimination of which is an animating purpose of leading proponents of a commons regime. It is widely considered desirable to preserve openness of the spectrum to a variety of viewpoints, sources of information, and technical spec-

---

elimination of the obsolete by permitting one another to use their proprietary intellectual property, while preserving the incentive for research and development activity by charging for such use. If computer manufacturer A has designed an improved screen, while its rival, B, has created a better memory, consumers will benefit if the two are allowed to meet and agree to permit one another to produce computers that share both improvements. Where the improvements involve many components with many proprietors, it has proven useful to create standard-setting organizations that decide upon compatibility standards that will enable the different innovations to work effectively together, without interference, and the antitrust authorities have not prevented such organizations from serving as forums in which rival firms can meet and decide upon the necessary steps for coordination and operability of the technical components.

Similar organizations can be useful in coordinating steps to control externalities such as interference in spectrum activities. They may constitute a practical step toward a regime such as Coase envisaged, in which property rights (such as are provided by patents for inventions) serve as the basis for mutually and socially advantageous agreements that can at least alleviate the interference problems. In sum, at least in this one respect, it is likely to be desirable to moderate the antitrust requirements as they are applied to spectrum use in the same way they have been moderated for intellectual property.

ifications. This subject is widely taken to be a critical matter and entered the debates over broadcasting regulation as early as the 1930s.[21] The FCC had taken upon itself to exercise some control over the content and sources of information transmitted over the spectrum, eliciting accusations of censorship and violation of the First Amendment. In particular, it was noted that those who defended such intervention advocated a treatment of spectrum activity that was different from that entailed in freedom of the press. Aside from steps to prevent the transmission of materials advocating hatred or mistreatment of minorities and sexually explicit materials, the FCC adopted rules such as the (since abandoned) "equal time" requirement, which was designed to ensure that no political candidate be precluded from broadcasting his message once a rival had been given the opportunity to do so. There have also been attempts to ensure increased children's programming. Moreover, the establishment of broadcasting by public television and public radio was evidently justified in part by a desire for preservation of diversity in programming, based on the belief that market forces would lead broadcasts to focus largely or even exclusively on programs constituting "the lowest common denominator."

This is not the place to discuss the pros and cons of the concern for diversity nor to examine the grounds for differentiation of the treatment accorded to newspapers and broadcasts in this arena. It is sufficient to recognize that in a democratic society, there is a presumption against control of all programming content by the government, particularly when that content can be deemed

21. For an illuminating account, see Coase (1959, p. 7).

to constitute propaganda, and with opposition viewpoints denied access to all broadcasting facilities. It would surely be equally suspect if a private individual or group were to obtain control of all or even the vast preponderance of broadcast facilities and exercised this control to exclude dissenting opinions.

The issue is evidently not a matter of economics, and economists cannot claim any professional qualification to opine on the best ways to deal with the matter. But two observations related to the economics are pertinent. First, it would seem that measures for the prevention or control of monopoly power, such as have already been referred to here, would also help to prevent foreclosure of diversity. Thus limitation of the share of spectrum activity that is permitted to any one individual or group will not only help prevent acquisition of monopoly power but also help preserve the opportunity for entry by others with divergent viewpoints. It is noteworthy that in the debate over freedom of the press versus FCC limitations on broadcast content, an argument that was offered in defense of this differentiated treatment was that entry into newspaper publication was easy (and at the time presumably less costly), so that the free market could be relied upon to ensure freedom for expression of heterodox viewpoints in the press.

Second, although it may be tempting to argue that a commons regime will provide the desired guarantees for freedom for diversity, this is not so, for reasons already discussed. Because spectrum usages are likely to entail scale economies and network externalities, it does not follow that the absence of regulation-imposed entry barriers will effectively open the gates to a multiplicity of independent and competing suppliers of spectrum services. Where scale economies or network externalities are present, market forces will make for the survival of a single supplier or at most a small number of suppliers of spectrum services, and with no further steps to prevent the problem, such freedom of entry will do little to preserve diversity.

Broadband Service to Rural Areas

Providing telecommunications service *universally* to all interested consumers has long been a goal of U.S. telecommunications policy. The issue arises because the cost per use or per user of deploying a telecommunications network is heavily dependent on population density: rural areas, being less densely populated than urban areas, are more expensive to serve on a per customer basis, and therefore investors tend to shy away from them. Historically, telecommunications carriers have provided plain old telephone service (or POTS) to high-cost (primarily rural) areas through a system of cross-subsidies encouraged by regulators.[22] This has enabled rural customers to receive service comparable in quality and price to that provided to customers in lower-cost urban and suburban areas.

22. In telecommunications the cross-subsidies were provided by AT&T, once the monopoly carrier. The firm was permitted to make up for any losses on the sparse routes by higher charges on the crowded routes. But competition generally impedes or altogether prevents such recoupment of the losses because entrants can be expected to flock to the overpriced services (they engage in "cream skimming"). For an arrangement that permits the preservation of a cross-subsidy desired by regulators even though entry has occurred, see Baumol (1999).

More recently, policymakers have indicated concern that *broadband* services are less accessible in rural areas. There are differing views as to the appropriateness of treating access to broadband as a universal service—that is, a service that should be equally available to all as a matter of policy. However, it is clear that in practice its deployment has been skewed away from rural areas. And upgrading of wireline networks to enable them to provide broadband services in low-density areas (or deployment of new networks capable of providing such services) is likely to be expensive and have a long payback period.

Some have suggested that high-speed wireless networks can facilitate broadband penetration in rural locations because they are less expensive to deploy in low-density areas. Several technological options have been proposed for rural communities, including code division multiple access (CDMA), which operates in licensed bands of spectrum, and Wi-Fi, which operates in unlicensed spectrum bands.

For the purposes of this report, the relevant question is whether one of the two spectrum governance regimes under examination—a licensed (market) approach or a commons approach—would prove more helpful in achieving the goal of expediting broadband deployment in rural areas in a way that is socially beneficial. To answer that question definitively, one would have to have a detailed understanding of the economics and business process underlying individual broadband deployment decisions. However, what we can say is that access to licensed spectrum typically represents a significant cost of doing business—one that (as with any business cost) must be covered by projected revenues if a firm is to proceed with the proposed deployment of spectrum services. The fact that service providers operating in unlicensed spectrum do not pay for access to that spectrum is no doubt a major reason that the commons approach appears to offer greater promise for the general welfare because it would seem to facilitate provision of broadband service to rural areas. But that is an illusion if added spectrum use and its crowding effect entail a cost to society, as is all too likely. In that situation, access is not *really* free, and failure to charge for spectrum access constitutes a subsidy to the investor that has merely been disguised, not eliminated. In contrast, in a market-based regime, in which entrepreneurs bid for the right to use spectrum, the price of the spectrum will reflect any scarcity and crowding effects. That is, if bidding arrangements are competitive, the price paid by the investor will correspond to the true cost to society, as is evidently appropriate.

Moreover, casual review of developments indicates that despite the cost of licenses, many wireless operators have acquired them, made investments that enable them to provide voice service and other service offerings to rural areas using licensed spectrum, and plan to continue to invest. Some of these investors, it is true, are already recipients of subsidy from the Universal Service Fund. But, in addition, the market regime itself has a mechanism that mitigates the cost of entry into low-usage areas. Specifically, wireless operators have been able to acquire licenses to serve rural areas at relatively low prices exactly because those prices reflect the lower profits that such service can be expected to yield. Stated differently, spectrum-use licenses to serve rural areas command a low price because the spectrum is not scarce.

We must, however, recognize that there may be special cases in which—because of very low density levels—even the most innovative business plan to provide rural service cannot produce a positive return. If provision of broadband services to rural areas is considered an imperative social good, a subsidy may well be appropriate (and necessary) to accomplish this goal. But access to the spectrum by special immunity from a license fee is just a way to conceal this subsidy; it is not a way that eliminates the true social cost, making it disappear by an act of magic. Economic rationality indicates that if policy dictates the use of subsidies, then such a program of subsidies should be clear and explicit rather than implicitly concealed.

Vested Interests versus Adaptable Arrangements

We turn last and hardly least to a critical issue for efficient spectrum governance in the long run. As has already been said, one of the primary shortcomings of the current spectrum regulation regime has been its failure to adapt to technological developments. There is no doubt a good deal of substance to the assertions about the increased availability of spectrum capacity, a development that could not have been accurately foreseen in the first half of the twentieth century, when the current regime evolved. But there is no reason to believe that we have since that time become much better at prophecy.

History provides abundant illustrations of almost bizarre misreadings of the future by experts in their fields. There is, of course, the famous tale of the director of the patent office who proposed that the office be terminated because the future did not offer the prospect of much further addition to the stock of technology. One can easily produce a series of quotations from the U.S. Geological Survey, the Department of the Interior, and the U.S. Bureau of Mines, beginning in the nineteenth century, each predicting imminent exhaustion of the supply of petroleum, each followed soon after by an increase in the size of proven reserves.[23] Then there is the famous 1943 quote (see chapter 1) attributed to Thomas Watson Sr., chairman of IBM, predicting a five-computer future market.

What all this implies is that even if at this moment there is an abundance of underutilized spectrum capacity, we have no way of knowing whether this will continue, whether the situation will improve even further, or whether it will deteriorate drastically.[24] One can envision the possible arrival of some new techniques and activities, whose uses and requirements are beyond our imagination, that become a matter of great priority for the general welfare, but whose operation requires immense amounts of spectrum capacity, meaning that they threaten drastic interference with other spectrum activities. We are not suggesting that it is rational to devote enormous concern to this perhaps remote possibility. But the key implication is that while any particular type of change cannot be foreseen, the prospect that some drastically different

23. See Baumol and Blinder (2005, p. 430).

24. As one recent example of government's poor powers to forecast spectrum needs, in 1994 the FCC allocated spectrum based on a projection that there would be 54 million users of domestic mobile services by the year 2000. In fact, users of such services numbered approximately 110 million in 2000— roughly double the commission's six-year-out forecast. FCC (2002, p. 12).

technology with unforeseeable spectrum requirements will appear is a virtual certainty. And this is a certainty for which rationality requires us to prepare.

It is also important here to draw attention to a mistaken inference into which one is all too easily misled. It is simply not true that crowding problems are always reduced by technological progress or that technological progress will continue to reduce crowding, even if it has done so in the past. Indeed, the problem can, sometimes unexpectedly, be seriously exacerbated by such advance. A clear example is the crowding of the skies and the associated delays of landings and takeoffs that could hardly have been foreseen by the Wright brothers on the day of their triumph at Kitty Hawk. But today a major problem facing competition among the airlines is shortage of landing slots, and the role of the flight controller has become indispensable for preventing frequent catastrophe in the crowded skies.

Since we cannot foresee the character of such developments, we have no way of dealing with them by instituting rules today that will suit their requirements tomorrow. But what we can and must do is to build explicitly into the spectrum management regime flexibility that permits the rules to be adapted to the new circumstances, as the character and requirements of those changes become clear, and to do so promptly and effectively. That is the true lesson of the generally acknowledged shortcomings of the current arrangements, stemming from the inflexibility of the obsolete regulations adopted the better part of a century ago.[25]

25. The discussion of Faulhaber and Farber is noteworthy, among other reasons, because it explicitly addresses the difficulties of transition from the

The primary source of resistance to change in any regime lies in the vested interests that it creates. Illustrations are easy to find, and their lessons are obvious. In New York City, for example, there is an enduring shortage of taxicabs because of the limited number of operator licenses ("medallions") that the current rules allow. But any attempt to expand the number of medallions is, predictably, met by the determined opposition of the current holders, the value of whose investment in those medallions would thereby be sharply reduced. Nor is their resistance to the socially desirable change unjustifiable. Because of the scarcity of medallions, the amounts they were forced to pay for them in the open market were already exorbitant. They have good grounds for arguing that the issuance of more medallions would constitute an illegal taking. The current medallion holders' vested interest becomes the bastion of opposition to socially desirable change, and rational planning should have sought in the first place to prevent this interest from arising.

This example illustrates a fundamental problem in designing a regime of spectrum management capable of meeting the needs of the future: it must avoid creating powerful, vested interests that will vigorously resist any change in the current arrangement, whatever its damage to the public interest. Perhaps the most effective of the imperfect solutions that seem to be available is a modified market arrangement characterized by spectrum licenses of

---

current arrangements (2002, p. 27). That is, they recognize that one of the basic impediments to progress in rectifying the shortcomings of the current regime is the resistance to change that is built into it. The importance of adaptability to change is also addressed in Faulhaber's article on spectrum policy (forthcoming).

unlimited duration but with federal intervention permissible in two types of situations where the pure property rights approach advocated by some analysts (private ownership of spectrum with no regulatory oversight) can damage economic efficiency:[26]

—On an incremental basis, the pertinent regulatory agency should be able to tighten regulatory limits on externality interference—for example, in response to continued improvements in transmitter filtering technology. Under a pure property rights approach, licensees would in some circumstances be induced automatically to achieve such changes via negotiation between the parties that create interference and those that are damaged by it. However, as Coase himself emphasized, this is unlikely to work where the number of parties involved is large and particularly when the source of the imposition is not readily determined nor the magnitude of the interference easily and accurately measured. Moreover, even where a property rights regime could perform the task effectively, the imposition of a change in the regulation is likely to reduce transaction costs and thereby help commoditize spectrum—an improvement in market efficiency that individual licensees may not appreciate.[27] The FCC should also provide incentives for adoption of better *receiver* filtering technology over

26. Several economists recently have called for a system in which "neighborly bargaining," supplemented by use of the courts to resolve remaining disputes, would eliminate the need for a federal spectrum regulator altogether. See Hazlett (2001, pp. 460–63) and Faulhaber (forthcoming). Under these approaches, the FCC would establish the initial license conditions, including limits on power spillage into adjacent bands and locations. Neighboring licensees could then change those limits through mutual agreement.

27. See Kwerel and Williams (2002, pp. 44–48) for an excellent discussion of interference control under a flexible licensing regime.

time, as the cost of filtering falls and the value of spectrum rises. This would permit the FCC to relax highly restrictive in-band power limits.

—The regulator should be able to withdraw licenses from incumbents on an exceptional basis, where technological developments or other unforeseen circumstances call for a fundamental restructuring of spectrum rights that the market alone will not achieve or only achieve with socially costly delays and market imperfections. In exchange, incumbents can be given transferable vouchers equal to the market value of the spectrum rights they relinquish. In such a regime, incumbents can be allowed to keep the proceeds from the sale of their licenses or buy back their modified licenses at no net cost.[28] Granted, under a pure property rights approach, the federal government could achieve a similar objective through exercise of the power of eminent domain. However, vested interests can delay, impede, or even block the government's use of eminent domain indefinitely through litigation. In contrast, if licenses are subject to the explicit threat of withdrawal, and licensees have assurance they will be made whole, the likelihood of effective and prompt adaptation to changing and unforeseen circumstances is materially increased.

Advocates of pure property rights in spectrum may fault this approach for preserving an explicit role for a spectrum regulator: in their view, it is necessary to eliminate the regulator in order to

28. Kwerel (2004) has proposed the use of vouchers in another context—namely, as part of a mandatory scheme to free up spectrum in the 300 MHz–3000 MHz range for market allocation. His proposal is a variation on the voluntary "big bang" transition scheme proposed by Kwerel and Williams (2002).

eliminate the attendant opportunities for rent seeking. But the federal government almost certainly will continue to maintain some regulatory oversight of spectrum—among other reasons, to protect national security and facilitate dispute resolution. Thus the goal of total elimination of spectrum regulation seems unrealistic and is probably undesirable as well. Moreover, the primary key to elimination of rent seeking is replacement of regulatory allocation of spectrum licenses with a market-based allocation, as this analysis advocates. Although the modifications proposed here to a purely market-based allocation may provide limited opportunities for regulatory politicization, that risk seems a worthwhile price to pay for a system sufficiently adaptable to permit ready accommodation to unforeseen developments.

Other arrangements that provide future flexibility are also possible.[29] One alternative approach is a temporary licensed property arrangement. Specifically, licenses would be valid for a limited period whose duration was clearly specified. The period must not be negligible in length, perhaps on the order of ten to fifteen years, so as to preserve investment incentives, without ossification of current ownership patterns and without creating powerful vested interests that can prevent necessary changes. In addition,

29. Noam (1998) suggests an interesting alternative: charges for a spectrum activity when it uses a portion of the spectrum that is crowded, such that the greater the demand for the use of a parcel of spectrum at any given moment, the higher the charge. This might well be an efficient way to control the externality-interference problem, but there are evident difficulties in defining or measuring of crowding and of metering the use that an activity makes of a crowded portion of the spectrum, as noted above. Moreover, it is not clear that interference, the real problem at issue, is proportionate to observable crowding, however defined.

the licenses would be obtainable *and renewable* only via an unbiased auction, so that the current holders could not simply presume that the licenses they now hold would automatically be reassigned to them, though perhaps some limited advantage could be offered to current holders as an incentive for investments with longer-term payoffs.

There are two drawbacks to this approach, however. The first is the effect upon incentives for investment. Granted, a ten- to fifteen-year horizon is of significant duration in a modern high-tech activity where major changes in models, products, and activities are constant and life cycles are very brief. Nevertheless, licensees would face a disincentive to investment as they reached the end of their lease.[30] Second, under this approach, licenses would expire and come up for reauction in a gradual fashion. However, the fundamental restructuring of spectrum rights envisioned above is likely to require a "big bang" type of transaction in which a large amount of interdependent spectrum could be put up for bid at the same time.

As a third possible drawback, some analysts have argued that letting spectrum rights revert back to the government periodi-

30. Under some circumstances, one could work around an end-of-lease investment problem. For example, a tenant with a long-term lease on a home or office could renegotiate the lease before undertaking a major renovation. However, such a renegotiation would not be possible with limited-duration spectrum licenses because of the corollary requirement that the licenses be reauctioned upon expiration. Alternatively, limited-duration licenses could work if parcels of spectrum were to become fungible, so that a licensee could easily move its transmitters and receivers from one frequency band to another. Although future technologies may well make spectrum more interchangeable, at present a licensee's capital investments are tied to specific frequencies.

cally runs the risk that regulators will delay reauctions or in other ways allow politics to trump economic efficiency. However, this argument is not highly persuasive. The federal government routinely reauctions licenses for the (temporary) rights to natural gas, timber, and other resources. Similarly, it routinely makes and carries out other commitments—for example, as to the duration of patent rights. Thus it is not unreasonable to assume that the pertinent federal agency will adhere to a commitment to hold a future reauction.

It may be thought that a commons regime is an attractive alternative—a good way to ensure the desired flexibility. But that is a misunderstanding. After all, those who take advantage of the opportunity to make use of a common property will be no more willing to give up that use voluntarily than if they had been required to pay for the privilege. The experience from the reign of Henry VIII to the eighteenth century of the closure movement, undertaken to close down the medieval commons, surely shows that this arrangement is no way to avoid the creation of vested interests. Faced with the loss of their means of livelihood, poor farmers and laborers put up a ferocious resistance to the closing of the commons, a response that continues to be noted by historians.[31]

31. The English closure (or enclosure) movement, which led to the conversion of communally owned or controlled property to private ownership, had a profound social and economic impact. Most economic historians believe that the movement had desirable long-term results, raising England's agricultural surplus and productivity by transferring poorly managed common property into the hands of individual owners who had an incentive to invest (for example, in crop rotation or a drainage system) and to avoid ecological damage. However, it resulted in the dislocation of large numbers of

Nor is this only a matter of ancient history. One need merely recall the resistance of the war veterans to the closing of their shantytown in Washington, D.C., during the Great Depression or the difficulty of evicting squatters from abandoned buildings to recognize that even an illegal commons creates vested interests that will vehemently oppose change. Should there be a recurrence of spectrum crowding problems, a commons, once created, can confidently be expected to be the focus of opposition to the needed changes.

Moreover, such opposition is likely to succeed in considerable part because of the direct government controls and regulations that a commons regime would necessarily entail, as discussed earlier. In addition to its inherent inefficiency, regulation provides countless opportunities for opponents of change to use the courts, Congress, and the administrative process to engage in tactical delay and to distort the policy process in their favor. Thus it provides the ideal terrain on which vested interests can defend the status quo and resist the efforts of those who seek to impose a new order.

A current example involves the large swath (400 MHz) of choice spectrum that is assigned to television broadcasting. Broadcasters occupy more than double the amount of spectrum that has been assigned to mobile voice and data services, despite the fact that nearly all Americans now receive their television signals through cable TV, direct broadcast satellite, and other nonbroadcast technology. The FCC formally reallocated a quarter of this

---

poor farmers and laborers, who had regarded the land as "theirs and their heirs'." See, for example, McCloskey (1972).

spectrum for flexible use, but broadcasters have been able to use their clout in Congress to block this change, at an annual opportunity cost to consumers of tens of billions of dollars.[32]

Summary: Superiority of Tradable Licenses to a Commons Regime

It is easy to understand the lure and popularity of a commons approach. The terms "sharing" and "cooperation" are readily suggested by the idea. More than that, the prospect of getting something for nothing—access at a zero price, with exclusion of no one who desires entry—has an irresistible lure.

But from the standpoint of the general welfare, access to common property is seldom truly free. As economists argue, all inputs are scarce—none of them has infinite capacity, and as such, they have what in economic jargon is described as opportunity cost. That is, use of such a resource for one purpose restricts or interferes with its use for other purposes. And so long as excess capacity is not abundant and readily available, a zero price for use of the resource is a license for overuse and needless interference, with

32. Hazlett (2005, pp. 249–51). According to Hazlett, the value to consumers of "broadcasting" (as opposed to broadcast content) is, at most, equal to the cost of delivering alternate service to the 10 million households that still receive only over-the-air TV signals. By contrast, American consumers receive more than $80 billion *a year* in benefits from the comparatively small amount (180 MHz) of spectrum allocated to wireless telephony—the spectrum service that is most free of FCC restrictions on use and resale. Thus he estimates that a substantial reallocation of spectrum from broadcasting to wireless telephony would generate consumer gains of two orders of magnitude or more.

haphazard and irrational selection of the users who enjoy the benefits and create the problems.

This, the tragedy of the commons, is widely recognized, and observers have understood its basis—that anything that belongs to everyone really belongs to no one. As is shown by the quotation at the beginning of this report, the problems raised by a commons have long been understood, though there are those who have equally long chosen to ignore them.[33]

Evidently the commons approach offers no guaranteed solution to any of the major problems of spectrum use that have been discussed here. That a commons exacerbates any problems of overcrowding and interference and impedes their solution is well known. Direct government controls can help to alleviate the problem, but only at a substantial real cost to society and to those directly involved.

Nor would a commons regime promote innovation, primarily because unlicensed entry removes the principal incentive for

33. Discussions of the issue predate Mrs. Marcet's writings. The issue was raised, for example, by Adam Smith in his university lectures, which preceded the *Wealth of Nations,* though his focus is the tragedy of the anticommons, the possibility that prospective users of resources will unnecessarily be precluded from their use: "The fish of the sea and rivers are naturally common to all; but the encroaching spirit that appropriated the game to the king and his nobles extended to the fishes . . . in the same manner, the water of rivers and the navigation of them, the navigation or right of sailing on the sea, is common to all. No one is injured by such use being made of them by another." See Adam Smith, *Lectures on Jurisprudence: Report of 1762–3,* in Meek, Raphael, and Stein (1982, p. 246). Note that this is a clear example of the impossibility of perfect foresight—Smith evidently did not foresee the possibility of overfishing and depletion of fish stock or the possibility of overcrowding of waterways (or the skies) by increasing traffic.

investment and risk taking by the users, who lose the assurance that they will be the beneficiaries. It is true that, as others have suggested, it will be in the interest of manufacturers of equipment for spectrum-using activities to invest in the design of technological solutions to the interference problems, as a means for the expansion of their volume of business. But as has been shown earlier, these discussions overlook the fact that this approach entails a fundamental externality problem of its own. So it will pay any such manufacturer to spend much less on research and development than the public interest requires. Equipment manufacturers simply cannot be expected to devote enough resources to deal effectively with any problems deriving from spectrum crowding.

The commons can conceivably do better at containing monopoly power and ensuring diversity, as its proponents maintain, but that is so only as long as there is no structural source of market power to impede entry or prevent it altogether. Since spectrum usages are likely to entail scale economies or network externalities, unlicensed entry may do little to contain monopoly power or preserve diversity in the absence of further remedies— remedies that would be equally effective if applied under a market approach.

Similarly, a commons regime may appear to offer a better approach for promotion of broadband service to rural areas because it makes spectrum available at no charge. But that ignores the almost certain cost to society of the added spectrum use and any resulting interference. The failure to charge for those costs represents a disguised subsidy, which undermines the price mechanism and its efficiencies, inviting misuse of spectrum resources.

Above all, a commons arrangement is an invitation to all comers to acquire positions of vested interest from which it will not be easy and perhaps not even feasible to dislodge them if and when currently unforeseeable changes in technology and other attendant circumstances make changes in regime an urgent matter.

4 The Possibility of a Mixed Regime

UP TO THIS POINT, the focus has largely been on two "pure" modes of operation: a commons approach and an approach resting on flexible licenses. But there are other arrangements that potentially can address some of the issues delineated here.

Privately Operated Minicommons

Other discussions have raised the possibility that enclaves of spectrum use will be created and operated voluntarily by private licensees in their pursuit of profit. There has been much discussion of the feasibility of such "commons parks" or "private commons" (and the FCC recently issued an order enabling licensees to establish "private commons").[1] However, it has been argued that they are unlikely and impractical because the transaction

1. FCC (2004).

costs would be too high and because the license holders would rather keep the available capacity for themselves, as indicated by the fact that nothing of this sort has so far been tried.[2]

Our own observations in analogous arenas suggest otherwise. If there is profit to be made from the charge of an entrance fee to such a park, then private enterprise and the profit motive can be relied upon to lead firms to carry out the necessary arrangements. And if entry into the commons is sufficiently beneficial to the entrants, there will indeed be profits to be made by giving them the opportunity to do so.

The analogy that seems most telling is provided by the licensing of intellectual property (IP).[3] One might well surmise that firms that hold patents on superior products or processes would determinedly seek to keep the technology to themselves and prevent others from using it. But a moment's thought indicates that this is not necessarily true. If another enterprise is willing to pay the IP proprietor more for being allowed to use it than the IP owner can obtain by using it exclusively by itself, then one can expect such sublicensing to materialize. And that is in fact what happens, as is documented in appendix A, which shows how voluntary granting of access to intellectual property, normally for an attractive financial return, far from being rare or nonexistent, is in fact commonplace. Moreover, it is no recent phenomenon but goes back at least to the last quarter of the nineteenth century.

The implication of this analogue is clear. If there is a financial benefit awaiting the prospective entrant into a commons park, the

2. Benkler (1998) and Werbach (2004).
3. Baumol (2002, especially chaps. 6, 7, and 13).

entrant will be prepared to pay a commensurate entry fee, and holders of the license that permits the setting up of such a park will be motivated to pursue the profits of its establishment and operation. Moreover, the license holder will profit by keeping the problem of interference under control to the degree called for by a benefit-cost calculation, for in that way the net benefit to the occupants of the park, and their willingness to pay for the right of entry, will be maximized. In practice, of course, this outcome can only be approximated, at best, but the point is not anything like a promise of perfection. Rather, it is to show that an arrangement based on flexible licensing of spectrum activity also offers the promise of some commons-like enclaves, as several proponents of a market approach have suggested.[4] More than that, this way of introducing a commons enclave will do so in a manner that preserves the incentives for efficiency of operation that only the market mechanism has been able to provide.

Easement versus Reliance on a Secondary Market

In addition to a privately run commons, which would be dedicated to use by unlicensed technology, it also may be desirable to have an arrangement that would allow new entrants to *share* (licensed) spectrum with licensees. A sharing arrangement is seen as providing a way for the new low-power or smart technologies to take advantage of spectral "white space."

Here, as with a dedicated commons, proponents of unlicensed access maintain that the market will not provide for such an

4. See, for example, Benjamin (2003) and Faulhaber and Farber (2002).

arrangement on its own because of high transaction costs and incumbent licensees' desire to protect their investment. In particular, Faulhaber and Farber have qualified their otherwise strong support for a market approach to spectrum governance by recommending that all property rights in spectrum entail a government-defined "noninterference easement," which other emitters could use on the condition that they not "meaningfully" interfere with the owner's right to transmit a clear broadcast.[5] This proposal evidently reflects their view that certain transmissions may have significant value as a secondary use but nevertheless may not be sufficiently valuable to justify the transaction costs of aggregating the necessary spectrum.

It is hard to disagree in principle with a proposal that, by definition, limits its scope to emissions that impose no costs on licensees. However, we are given to understand that there currently exist no types of equipment or techniques that can guarantee such a benign outcome, and pending such a development, for several reasons, the case for a government-defined easement remains questionable.

First, an FCC-imposed easement may not be needed if licensees and other parties can achieve the same end through reliance on a secondary market, as seems likely. Like IP owners, licensees holding flexible spectrum licenses will share their property if it is financially attractive to do so—for example, through an arrangement with one or more equipment vendors that provide access to the vendor's customers. And although sharing arrange-

5. "In effect, this easement creates a commons at all frequencies and in all locations of a special type: non-interfering uses only." Faulhaber and Farber (2002, p. 19). See also J. H. Snider (2005).

ments may entail higher transaction costs than a dedicated commons, one can easily imagine the emergence of streamlined mechanisms to handle those transactions, as has occurred in the IP
arena. Indeed, one can well be concerned that any governmental
easement arrangement will be subject to conditions and inspec-
tions to ensure that the promised degree of noninterference is not
exceeded, and experience surely indicates that the transaction costs
entailed in such a governmentally operated regime are unlikely to
be negligible.

Beyond being unnecessary, the creation of a government-
defined easement, by preemption of the development of a secondary market in so-called noninterfering emissions, may well
prove harmful. Moreover, users of spectrum on this basis may
acquire squatters' rights. Once the unlicensed users become accus-
tomed to having free access to licensed spectrum, they may use the
regulatory or political process to maintain that access, preventing
licensees from occupying more of their own spectrum.[6] The argument favoring market accommodation of such emissions rests on
the now familiar position that spectrum rights holders have an

6. The evolution of low power television (LPTV) illustrates the political
force of squatters' rights. The FCC created LPTV service in 1982 to provide
local and niche programming for rural Americans and other small, localized
groups of viewers. For many years, LPTV stations had "secondary" spectrum
status, which meant that they could not cause interference to the reception
of full-service TV stations and had to accept interference from such stations.
However, in 1999 Congress passed the Community Broadcasters Protection
Act, which gave qualifying LPTV stations "primary" status, including some
measure of protection from full-service stations, even as those stations con-
verted from analog to digital format. See FCC (2000). That upgrade in stand-
ing for LPTV apparently increased somewhat the technical difficulty of the
digital conversion process.

incentive to act in ways that result in (approximately) optimal use of spectrum space. There is reason to question whether a governmental rule without price and profit incentives will be able to match the performance of a market regime.

In sum, even those virtues that can be ascribed to a commons arrangement can probably be obtained, at least to a substantial degree, under a regime of flexible licenses such as is favored by the analysis here. The preponderance of the evidence seems to call for this approach as the most promising of the available alternatives in terms of its prospective contributions to the general welfare.

But there may be one noteworthy exception: there is much that may be gained from the reservation of very limited bands for experimental purposes. Experience indicates dramatically that the prospective uses and the identity of the potential beneficiaries of untried technical advances are virtually unpredictable. As a result, it is often very difficult to finance the testing of unproven inventions from which all of society stands to benefit. Consequently, there may be a justification for the government to establish experimental bands that would be available to meagerly financed inventors who could use these venues to test their technology. (In fact, the FCC currently grants temporary experimental licenses at no charge.) If the tests show such an innovation to be workable, then the owners of the intellectual property can be left to the market to obtain access to licensed spectrum on their own, on the basis of the proven technology. This is an arrangement that appears to deserve consideration and seems to be the most meritorious candidate for governmental provision of free and unlicensed access to the spectrum.

5 Conclusion: Tradable and Modifiable Licenses— the Most Promising of the Imperfect Solutions

A BALANCED DISCUSSION must avoid exaggerated claims for the advocated course of action. Here the conclusion offered is that the most rational way of dealing with the problems of spectrum governance is an arrangement using tradable licenses subject to limited regulatory intervention, where the uses of those licenses are flexible, and they can be rented to others, in whole or in part, for other uses. (For a review of the major options and issues discussed here, see appendix B.) Such an arrangement cannot be claimed to solve the problems of spectrum use with anything approximating perfection. Indeed, it is tempting here to recall the old remark, slightly modified, and suggest that the only things worse than the proposed arrangement are all of the alternatives.

To be specific, it should be clear that the approach proposed only deals with the problem of interference in a rough and ready way. Assignment of well-defined parcels of spectrum for use or resale only by the licensee directly limits crowding and, along with technical requirements for the performance of transmitters and receivers, indirectly limits interference. However, determination of the optimal requirements calls for information on costs and benefits that is generally not available to a government regulator, and evolving technology can make existing requirements obsolete. Negotiation between licensees can achieve some but not all of the desirable refinements in these requirements. In sum, the control of interference by exclusive assignment of spectrum-use licenses will be unavoidably imperfect and cannot make any claim to optimality.

The same is evidently all too true of reliance on the antitrust and regulatory rules to contain monopoly. There is a voluminous and authoritative literature on the imperfections of these institutions. Still less can be said about the protection of diversity in spectrum usage. As discussed earlier, rules limiting the share of the available spectrum that can be acquired by any one party can help. Direct forms of intervention for this purpose, such as equal time requirements, have been tried and found wanting, though no good substitutes have been found. Financing of public television and radio by government does help, but it hardly commands universal approbation. Technical progress can help alleviate the diversity problem, as when the advent of cable television multiplied the number of available channels more than tenfold. One can only conclude that *none* of the contemplated spectrum governance arrangements deals effectively with the problem of protecting diversity.

Ultimately the superiority that can be claimed for a flexible license regime rests on three observations. The first is that it entails a straightforward modification of the current and hence familiar regime, and should therefore minimize transition problems. Second, it takes maximal advantage of the well-documented economic efficiency properties of the market mechanism, tending to allocate resources to those who can make most effective use of them and ensuring, in the presence of competition, that the resources are not used wastefully. In particular, the flexibility of the arrangement provides a powerful disincentive for spectrum hoarding, to which informed observers have attributed much of the phenomenon of spectrum crowding. The third and most important argument favoring the flexible license regime is its straightforward adaptability, the ease with which it can be modified to accommodate unpredicted and unpredictable changes in the nature of spectrum utilization, changes that will certainly occur.

Appendix A:

An Analog Suggesting Viability of a Private Commons: Voluntary Licensing of Intellectual Property

ONE CAN READILY document the existence and profusion of markets in intellectual property, as well as their substantial history. In several articles, Lamoreaux and Sokoloff have provided striking evidence about the early beginnings of markets in technology and their subsequent growth. By painstaking study of patent records and related materials, they have been able to document the transactions.[1] They report that "such exchange began to take off by the middle of the nineteenth century." New information channels and intermediaries began to appear at this early date, with specialized patent agents and lawyers advertising their availability as finders of buyers of patents on a commission basis. "Their numbers began to mushroom in the 1840s, first in the vicinity of Washington and then in other urban centers." Particularly with a secular increase in the percentage of inventions contributed by

1. Lamoreaux and Sokoloff (1996).

individuals who specialized in the activity, rather than by part-time inventors, "trade in patent rights increased in all regions [of the United States] through 1910 [the date at which the statistical analysis terminates], nearly doubling overall [from 1870–71] by this measure." The authors conclude that "the growth of intensely competitive national product markets, coupled with the existence of the patent system, created a powerful incentive for firms to become more active participants in the market for technology."[2]

Today, the sale, licensing, and trading of technology has become a large-scale activity. Arora, Fosfuri, and Gambardella list a sample of "leading deal makers in markets for technology" that includes companies such as Microsoft, IBM, AT&T, Monsanto, Motorola, Bell South, Daimler-Benz, Eli Lilly, Eastman Kodak, Sprint, Philips Electronics, Siemens, General Motors, Honeywell, Boeing, Fiat, Ford, General Electric, Hitachi, Toshiba, Dow Chemical, Johnson and Johnson, and many others.[3] They report the results of a survey of 133 companies by a British consulting firm, which indicated that 77 percent of the companies studied had licensed technology from others while 62 percent had licensed technology to others. However, they report that in terms of the budgets involved, licensing is a fairly modest activity compared to internal research and development. The survey estimates that expenditures to license technology from others amount to 12, 5, and 10 percent of the total research and development budgets of respondents from North America, Europe, and Japan, respectively. However, they estimate that the size of the "market for

2. Lamoreaux and Sokoloff (1996, pp. 12687, 12689, and 12691).
3. Arora, Fosfuri, and Gambardella (2001, pp. 34–37).

technology" is about $25 billion in North America alone, which, they note, is about the size of the 1996 gross domestic expenditure on research and development in France and greater than that of the United Kingdom.[4]

For many firms, participation in technology markets is of critical importance. For example, selling access to polypropylene technology has constituted a major activity of the Union Carbide Corporation, and IBM has informed us that it has a technology exchange contract with every major manufacturer of every significant computer part throughout the world. We have data that enabled us to estimate that in the year 2000, approximately 20 percent of IBM's total profits derived from the sale of licenses.[5]

It is clear that voluntary dissemination is no isolated and unusual phenomenon. The Licensing Executives Society reports a membership of nearly 10,000 from more than sixty countries, and it runs seminars and conferences such as one on "Leveraging Technology for Competitive Advantage." There are many websites offering information and help for licensing and technology transfer. According to the U.S. National Science Board, between 1980 and 1998, American, European, and Japanese firms arranged some 9,000 strategic technology alliances.

4. Arora, Fosfuri, and Gambardella (2001, pp. 30–31).
5. Baumol (2002, p. 84, footnote 2).

Appendix B:

Summary

THE MAIN REQUIREMENT for efficient governance of the radio frequency spectrum is that it be fully and easily adaptable to unforeseeable changes in the needs of the public interest as technology, spectrum-using services, and the public's demands evolve. Impediments to such adaptability account for the inefficiency and waste inherent in the Federal Communication Commission's traditional administrative approach to regulating the spectrum. Under this approach spectrum licenses, designed to limit interference, were awarded through a slow, costly, and obsolete process, and rigid restrictions on the use and transferability of licenses kept valuable bandwidth locked into the provision of low-value services and encouraged hoarding of spectrum by licensees. Although the FCC is gradually dismantling its pervasive regime of command-and-control regulation, neither a pure market nor a commons regime—the ameliorative approaches now most widely advocated—avoids the creation of vested interests that can be counted on to resist appropriate adaptation to evolving circumstances.

The Basic Problem: Crowding and Interference

The essence of the spectrum usage problem is the phenomenon of crowding, which arises because the spectrum appears to remain a scarce resource and is therefore subject to costly mutual interference by the profusion of users. As is well known, where the number of activities and affected individuals is large, the market by itself will not lead to an efficient allocation of a scarce resource that is subject to inadvertently caused damage, such as interference.

Amelioration via Transferable Licenses

Under the transferable license approach to spectrum governance, which is advocated here, licenses with flexibility of use are subject to resale or lease to others and are initially granted via a market process such as auctioning. This arrangement can deal with the crowding problem and can readily be adapted to changing circumstances. The FCC has embraced auctioning of new licenses and begun to institute other decentralized market mechanisms. Although these reforms have led to measurable gains in efficiency and investment, administrative restrictions still cover much of the spectrum, and FCC decisions remain subject to political pressure from entrenched interests.

Commons Regime as an Alternative

Recently, as an alternative approach to spectrum governance, it has been proposed that some or all of the spectrum be treated as a common resource that would be open to all entrants. Proponents of this commons approach maintain that with new interference-avoiding technology, licensing is becoming both unnecessary and impractical. That confident claim seems questionable, even for the present. But even if not immediately, it is surely unsustainable for the future. After all, who could have foreseen soon after the early automobiles made their appearance that traffic jams and crowded highways would become a hallmark of transportation in the future? However, the prospect that serious crowding problems will persist or reappear is less of a concern than the certainty that the commons procedure will create the powerful vested interests that preclude appropriate adaptation to changing circumstances.

Issues That Must Be Addressed

There are six key issues with which a rational spectrum regime must deal: interference, adequacy of investment in innovation, monopoly power, preservation of diversity, service to rural areas, and the tension between vested interests and the need for adaptable arrangements. The tradable license approach and the commons approach, as well as two proposals for a mixed approach (a "private commons" and an "easement" in licensed spectrum for "noninterfering" uses), have been evaluated here in terms of how they address these six issues.

Benefits of a Modified Market Regime

This discussion concludes that the most rational approach to spectrum governance entails auctioned licenses whose uses are flexible and that can be rented to others, in whole or in part, for other applications. Regulatory intervention should be permissible in a limited number of situations—primarily, the exceptional case where technological developments or other unforeseen circumstances call for a fundamental restructuring of spectrum rights that the market alone will not achieve or achieve only with socially costly delays and market imperfections. Although this variation on the pure market approach does not solve the problems of spectrum use with anything approximating perfection, it offers three major advantages:

—It entails a straightforward modification of the current regime and should therefore minimize transition problems.

—It takes maximal advantage of the efficiency properties of the market mechanism, tending to allocate resources to those who can make the most effective use of them and ensuring, in the presence of competition, that resources are not used wastefully.

—Most important is the ease with which it can be modified to accommodate unpredicted and unpredictable changes in the nature of spectrum utilization.

Attractions of a Commons Regime and Its Shortcomings

It is easy to understand the lure and popularity of the commons approach as an idea that promises something for nothing—access at a zero price, with exclusion of no one who desires entry. But from the standpoint of the general welfare, access to common property is seldom truly free. So long as excess capacity is not abundant and readily available, a zero price is a license for overuse and needless interference, with haphazard and irrational selection of the users who enjoy the benefits and create the problems.

Claims that the Crowding Problem Is Now Solved versus Unpredictability of the Future

Although proponents of a commons approach maintain that new technology is overcoming the problem of interference, experience suggests otherwise. FCC efforts to accommodate new and better unlicensed wireless technology, by designating certain bands for unlicensed use, repeatedly have resulted in overcrowding. New technology has expanded the amount of spectrum capacity that is effectively available in these bands, but demand has more than kept pace, and this is unlikely to change. We all know from bitter experience that overcrowded roads generally continue to be overcrowded and only unclog when a system of tolls or some other rationing device is put into place. Spectrum is not essentially different.

Need for Direct Controls in a Commons Regime

Precisely because technology alone cannot be relied upon to prevent overcrowding, proposals for a commons approach envisage reliance on protocol requirements and other direct government controls. However, direct controls tend to be inefficient and costly, and they create an incentive for rent seeking. Predictably, in response to interference in the unlicensed bands, some wireless Internet service providers are seeking exclusive rights within the commons—a contradiction in terms.

Incentives for Innovation under Unlicensed Entry

Unlicensed entry provides no particular advantage as far as promoting innovation. The profit mechanism has produced an outpouring of innovation unparalleled in previous history. In the case of spectrum, if no one holds property (a license), that source of incentive for the cost, effort, and risk entailed in the innovation process will be undercut. Proponents of open access argue that manufacturers of interference-avoiding equipment would be as powerful an engine for innovation under a commons regime as license holders would be under a market regime. Although there is some validity to this contention, it overlooks the fact that if manufacturer A's investment in research and development succeeds in reducing interference problems, part of the resulting increase in business will go to manufacturers B, C, and D. Because much of the benefit will go to others, equipment manufacturers will not have the incentive to invest adequately in means to prevent interference.

A Commons Regime and the Issue of Monopoly Power

The commons approach can conceivably do better at containing monopoly power and ensuring diversity, as its proponents maintain, but that is so only as long as there is no structural source of market power to impede entry or prevent it altogether. Since spectrum usages are likely to entail scale economies or network externalities, unlicensed entry may do little to contain monopoly power or preserve diversity, in the absence of further remedies such as a simple ceiling on market share—remedies that would be equally effective if applied under a market approach.

Vested Interests in a Commons Regime and Resistance to Necessary Change

Finally, although a commons regime is often seen as a good way to ensure the desired flexibility, that is a misunderstanding. Those who take advantage of the opportunity to make use of a common property will be no more willing to give up that use voluntarily than if they had been required to pay for the privilege. Moreover, they are likely to succeed because a commons regime would necessarily entail a regulatory structure, and regulation provides countless opportunities for opponents of change to engage in tactical delay and distort the policy process in their favor. A current example of regulatory gamesmanship involves the large swath of choice spectrum assigned to television broadcasting. Although the FCC formally reallocated a quarter of that spectrum for flexible use, politically powerful broadcasters have been able to block the change, at an annual opportunity cost to American consumers of tens of billions of dollars. In short, a commons arrangement is an invitation to all comers to acquire positions of vested interest from which it will not be easy and perhaps not even feasible to dislodge them if and when currently unforeseeable developments in technology and other attendant circumstances make changes in regime an urgent matter.

References

Arora, Ashish, Andrea Fosfuri, and Alfonso Gambardella. 2001. *Markets for Technology: The Economics of Innovation and Corporate Strategy.* MIT Press.

Baumol, William J. 1999. "Having Your Cake: How to Preserve Universal-Service Cross Subsidies while Facilitating Competitive Entry." *Yale Journal on Regulation* 16, no. 1: 1–17.

———. 2002. *The Free-Market Innovation Machine: Analyzing the Growth Miracle of Capitalism.* Princeton University Press.

Baumol, William J., and Alan S. Blinder. 2005. *Economics: Principles and Policy.* 10th ed. Mason, Ohio: Thomson South-Western.

Benkler, Yochai. 1998. "Overcoming Agoraphobia: Building the Commons of the Digitally Networked Environment." *Harvard Journal of Law and Technology* 11, no. 2: 287–400.

———. 2002. "Some Economics of Wireless Communications." *Harvard Journal of Law and Technology* 16, no. 1: 25–83.

Benjamin, Stuart Minor. 2003. "Spectrum Abundance and the Choice between Private and Public Control." *New York University Law Review* 78: 2007–102.

Coase, Ronald H. 1959. "The Federal Communications Commission." *Journal of Law and Economics* 2, no. 1: 1–40.

———. 1960. "The Problem of Social Cost." *Journal of Law and Economics* 3, no. 1: 1–44.

De Vany, Arthur S., and others. 1969. "A Property System for Market Allocation of the Electromagnetic Spectrum: A Legal-Economic-Engineering Study." *Stanford Law Review* 21, no. 6: 1499–561.

Downs, Anthony. 1992. *Stuck in Traffic.* Brookings.

Faulhaber, Gerald R. Forthcoming. "The Question of Spectrum: Technology, Management, and Regime Change." *Journal on Telecommunications and High Technology Law.*

Faulhaber, Gerald R., and David J. Farber. 2002. "Spectrum Management: Property Rights, Markets, and the Commons." Working Paper 02-12. AEI-Brookings Joint Center for Regulatory Studies.

Federal Communications Commission. 2000. *In the Matter of Establishment of a Class A Television Service: Report and Order.* MM Docket 00-10 (April 4, 2000).

———. 2001. *Promoting Efficient Use of Spectrum through Elimination of Barriers to the Development of Secondary Markets: Comments of 37 Concerned Economists.* WT Docket 00-230 (February 7, 2001).

———. 2002. *Spectrum Policy Task Force Report,* ET Docket 02-135.

———. 2004. *Promoting Efficient Use of Spectrum through Elimination of Barriers to the Development of Secondary Markets: Second Report and Order, Order on Reconsideration, and Second Further Notice of Proposed Rulemaking.* WT Docket 00-230 (September 2, 2004).

Hazlett, Thomas W. 1990. "The Rationality of U.S. Regulation of the Broadcast Spectrum." *Journal of Law and Economics* 33, no. 1: 133–75.

———. 1998. "Assigning Property Rights to Radio Spectrum Users: Why Did FCC License Auctions Take 67 Years?" *Journal of Law and Economics* 41, no. 2 (part 2): 529–75.

———. 2001. "The Wireless Craze, the Unlimited Bandwidth Myth, the Spectrum Auction Faux Pas, and the Punchline to Ronald Coase's 'Big Joke': An Essay on Airwave Allocation Policy." *Harvard Journal of Law and Technology* 14, no. 2: 335–469.

———. 2005. "Spectrum Tragedies." *Yale Journal on Regulation* 22, no. 2: 242-74.

Kwerel, Evan. 2004. "Spectrum Exchanges and Incumbent Clearing." Slide presentation at Conference on Communications Policy. Stanford Institute for Economic Policy Research and AEI-Brookings Joint Center for Regulatory Studies, Stanford, Calif., October 9, 2004.

Kwerel, Evan and John Williams. 2002. "A Proposal for a Rapid Transition to Market Allocation of Spectrum." Working Paper 38. FCC Office of Plans and Policy (November 15).

Lamoreaux, Naomi R., and Kenneth L. Sokoloff. 1996. "Long-Term Change in the Organization of Inventive Activity." *Proceedings of the National Academy of Sciences* 93 (November): 12686–92.

Lessig, Lawrence. 2001. *The Future of Ideas: The Fate of the Commons in a Connected World.* Random House.

Levin, Harvey J. 1971. *The Invisible Resource: Use and Regulation of the Radio Spectrum.* Johns Hopkins University Press.

Marcet, Jane. 1819. *Conversations in Political Economy.* 3d ed. London: Longman, Hurst, Rees, Orme and Brown.

McCloskey, Donald N. 1972. "The Enclosure of Open Fields: Preface to a Study of Its Impact on the Efficiency of English Agriculture in the Eighteenth Century." *Journal of Economic History* 32, no.1: 15–35.

Meek, R. L., D. D. Raphael, and P. G. Stein, eds. 1982. *Adam Smith. Lectures on Jurisprudence: Report of 1762–3.* Indianapolis: Liberty Fund.

Minasian, Jora R. 1975. "Property Rights in Radiation: An Alternative Approach to Radio Frequency Allocation." *Journal of Law and Economics* 18, no. 1: 221–72.

Noam, Eli. 1998. "Spectrum Auctions: Yesterday's Heresy, Today's Orthodoxy, Tomorrow's Anachronism. Taking the Next Step to Open Spectrum Access." *Journal of Law and Economics* 41, no. 2 (part2): 765–90.

Rosston, Gregory L., and Jeffrey S. Steinberg. 1997. "Using Market-Based Spectrum Policy to Promote the Public Interest." *Federal Communications Law Journal* 50, no.1: 88–115.

Scrivner, John. 2004. *WISP Homestead Policy Proposal for WISP Use of the ITFS Band,* Comment filed with the Federal Communications Commission. WT 03-66 (March 19, 2004).

Shelanski, Howard, and Peter Huber. 1998. "Administrative Creation of Property Rights to Radio Spectrum." *Journal of Law and Economics* 41, no. 2 (part 2): 581–608.

Snider, J. H. 2005. "Reclaiming the Vast Wasteland: The Economic Case for Re-Allocating to Unlicensed Service the Unused Spectrum (White Space) between TV Channels 2 and 51." Washington: New America Foundation (October).

Spiller, Pablo T. and Carlo Cardilli. 1999. "Towards a Property Rights Approach to Communications Spectrum." *Yale Journal on Regulation* 16, no. 1: 53–84.

Werbach, Kevin. 2004. "Supercommons: Toward a Unified Theory of Wireless Communication." *Texas Law Review* 82, no. 4: 863–973.

White, Lawrence J. 2000. "'Propertyzing' the Electromagnetic Spectrum: Why It's Important and How to Begin." *Media Law and Policy* 9 (Fall).

Index

AT&T, 40n22
Auctions for awarding licenses and permits, 11, 22, 33, 50–51, 71–72

Benkler, Yochai, 7n1, 12n9

CDMA (code division multiple access), 41
Closure movement, 51
CMRS (commercial mobile radio services) licenses, 22–23
Coase, Ronald, 9, 19–20, 23, 47
Commons regime: adaptability of, 51–53, 77–78; direct controls advocated, 28–29, 76; efficiency of, 13–14; exclusivity of rights and, 13, 14; innovation investment, 32–34, 54–55; market approach vs., 10–11, 13; monopoly control, 14, 55, 77; politicization of the regulatory process, 29–30, 30–31; privately run, 57–59, 67–69; shortcomings of, 73, 75–78; technology component in,

12–13, 24–26; telecommunications service to rural areas, 41–42; tradable licenses vs., 53–56; traditional approach vs., 13
Crowding: causes of, 72, 75; hoarding and, 65; negotiated settlement approach to, 19–20; technology solutions, 27–29, 30–31, 44–45, 75; transferable license approach, 29, 72; transferable permit approach, 21–22; unlicensed use and, 18, 25–27, 41–42 . See also Interference control

De Vany, Arthur S., 23n5
Direct (government) controls, 28–30, 76
Diversity preservation, 14, 37–40, 55, 64, 77
Downs, Anthony, 27n13

Easements, imposed and noninterference, 60–63

JOINT CENTER